MATH
ADVENTURES

FUN WITH NUMBERS
AND
SHAPES

THE BOOK CONTAINS SOME OF THE BASIC AND ADVANCED CONCEPTS TO BOOST THE YOUNG BRAINS.

ER. ROUF

COPYRIGHT © 2024 BY ER. ROUF.

ALL RIGHTS RESERVED .

NO PART OF THIS BOOK MAY BE COPIED, REPRODUCED IN A RETRIEVAL SYSTEM OR TRANSMITTED IN ANY FORM OR BY ANY OTHER MEANS WITHOUT PRIOR WRITTEN PERMISSION OF THE PUBLISHER.

Contents:

1. BASICS OF NUMBER SYSTEM

- WHOLE NUMBERS
- NATURAL NUMBERS
- RATIONAL NUMBERS
- IRRATIONAL NUMBERS
- PRIME NUMBERS
- CO-PRIME NUMBERS
- COMPOSITE NUMBERS
- INTEGERS
- ODD AND EVEN NUMBER
- FACE VALUE AND PLACE VALUE

2. BASICS OF FRACTIONS AND IT'S TYPES

- PROPER FRACTIONS
- IMPROPER FRACTIONS
- LIKE FRACTIONS
- UNLIKE FRACTIONS
- MIXED FRACTIONS.

3. BASICS OF ADDITION AND SUBTRACTION

- ADDITION AND SUBTRACTION OF WHOLE NUMBERS

- ADDITION AND SUBTRACTION OF RATIONAL NUMBERS
- ADDITION AND SUBTRACTION OF IRRATIONAL NUMBERS

4. BASICS OF MULTIPLICATION

- MULTIPLICATION OF WHOLE NUMBERS
- MULTIPLICATION OF RATIONAL AND IRRATIONAL NUMBERS

5. BASICS OF ALGEBRAIC EXPRESSIONS

- VARIABLES, CONSTANTS, COEFFICIENTS AND
- FUNDAMENTAL OPERATOR
- ALGEBRAIC EXPRESSION IN ONE VARIABLE AND TWO VARIABLES

6. ALGEBRAIC EQUATIONS AND ITS TYPES

- LINEAR EQUATIONS IN ONE AND TWO VARIABLES
- QUADRATIC EQUATIONS AND CUBIC EQUATIONS

7. BASICS OF POLYNOMIAL

- MONOMIAL
- BINOMIAL
- TRINOMIAL

8. BASICS OF HCF AND LCM

- FACTORS AND COMMON FACTORS

- HCF AND LCM BY PRIME FACTORISATION

9. BASICS OF GEOMETRY

- SQUARE
- CUBE
- CONE
- CYLINDER
- CIRCLE
- TRIANGLE

10. SOME FUN FACTS ABOUT

- ZERO
- π
- INFINITY
- ONE (1)

WHOLE NUMBERS:

NUMBERS STARTING FROM ZERO (0) ARE CALLED WHOLE NUMBERS.

FOR EXAMPLE;

0, 1, 2, 3, 4, 5, 6……..AND SO ON.

NATURAL NUMBERS:

NUMBERS STARTING FROM 1 ARE CALLED NATURAL NUMBERS.

FOR EXAMPLE; 1, 2, 3 ,4, 5, 6…..AND SO ON

RATIONAL NUMBERS:

NUMBERS WHICH ARE WRITTEN IN THE FORM $\frac{P}{Q}$ where $Q \neq 0$ ARE CALLED AS RATIONAL NUMBERS.

FOR EXAMPLE;

$\frac{1}{2}, \frac{3}{5}, \frac{8}{9}, \frac{3}{1}, \frac{2}{1}$ etc

NOTE: **P** IS CALLED **NUMERATOR** & **Q** IS CALLED **DENOMINATOR**

OR

THE NUMBERS WHICH ARE TERMINATING AFTER THE DECIMAL POINT ARE ALSO TERMED AS RATIONAL NUMBERS.

FOR EXAMPLE;

1.5, 3.32, 5.67, 0.45, ETC.

SINCE AFTER THE DECIMAL POINT, THE EXPANSION TERMINATES.

IRRATIONAL NUMBERS:

NUMBERS WHICH CANNOT BE REPRESENTED IN THE FORM OF $\frac{P}{Q}$ ARE CALLED AS IRRATIONAL NUMBERS.

FOR EXAMPLE; √2, √3, √5…. AND SO ON.

ALL SUCH NUMBERS WITH UNDER ROOTS AND WHICH ARE NOT PERFECT SQUARES ARE TERMED AS IRRATIONAL NUMBERS .

OR

THE DECIMAL EXPANSION WHICH IS NON-TERMINATING AND NON-RECURRING OR NON-REPEATING ARE CALLED IRRATIONAL NUMBERS.

FOR EXAMPLE; 2.2346887486344…, 3.54676456…SO ON , AS THE DECIMAL EXPANSION NEITHER TERMINATES NOR REPEATS AFTER DECIMAL POINT.

PRIME NUMBERS:

NUMBERS HAVING ONLY TWO FACTORS, NUMBER ITSELF AND 1 ARE CALLED PRIME NUMBERS.

FOR EXAMPLE; 2, 5, 7, 11, AND SO ON.

$2 = 2 \times 1, 5 = 5 \times 1, 7 = 7 \times 1$ ETC .

NUMBERS ON LEFT SIDE 2, 5 , 7 ARE CALLED PRODUCT AND THE NUMBERS ON RIGHT SIDE WHICH ARE MULTIPLIED ARE CALLED THEIR FACTORS.

FOR EXAMPLE; 2 HAS ONLY TWO FACTORS, THE NUMBER ITSELF 2 AND 1, SAME IS THE CASE FOR 5, 7, 11, 13 AND SO ON.

NOTE: **1** IS NOT A PRIME NUMBER BECAUSE IT CANNOT BE FACTORED INTO TWO SEPARATE NUMBERS OR IT'S DIVISIBLE BY 1 MEANS ITSELF ONLY.

COPRIME NUMBERS:

NUMBERS HAVING HCF **1** ARE CALLED AS CO-PRIME NUMBERS.

FOR EXAMPLE; (3,5), (7,9) ETC.

WE WILL LEARN TO CALCULATE HCF IN NEXT CHAPTER.

COMPOSITE NUMBERS:

NUMBERS HAVING MORE THAN TWO FACTORS ARE CALLED COMPOSITE NUMBERS. **FOR EXAMPLE**; 4, 8, 9, ETC

$4 = 2 \times 2 \; also \; 4 = 4 \times 1$

SO 4 HAS THREE FACTORS 2, 4, 1, SO 4 IS A COMPOSITE NUMBERS.

EVEN NUMBER:

NUMBER WHICH IS PERFECTLY DIVISIBLE BY 2 IS CALLED AND EVEN NUMBER.

FOR EXAMPLE; 2, 4, 6, 8, 10, 12 ETC

$\frac{4}{2} = 2, \frac{6}{2} = 3, \frac{8}{2} = 4$ *etc*

ODD NUMBER:

NUMBER WHICH IS NOT PERFECTLY DIVISIBLE BY 2 IS CALLED AN ODD NUMBER, HERE WE GET A DECIMAL AFTER DIVIDING BY 2.

FOR EXAMPLE; 1, 3, 5, 7, 9, 11, 13… ETC

$\frac{1}{2} = 0.5, \frac{3}{2} = 1.5, \frac{5}{2} = 3.5$ *etc*

FRACTION:

A FRACTION IS A WAY OF SHOWING A PART OF SOMETHING. IT HAS TWO NUMBERS **X** AND **Y** AND CAN BE WRITTEN IN THE FORM OF $\frac{X}{Y}$

X IS THE NUMBER CALLED **NUMERATOR** WHICH IS ON TOP THAT SHOWS HOW MANY PARTS YOU HAVE.

Y IS THE NUMBER CALLED **DENOMINATOR** WHICH IS ON THE BOTTOM THAT SHOWS HOW MANY EQUAL PARTS THE WHOLE IS DIVIDED INTO.

TYPES OF FRACTIONS:

1. PROPER FRACTION:

THE NUMERATOR IS SMALLER THAN THE DENOMINATOR AS

$X < Y$

FOR EXAMPLE; $\frac{3}{4}, 3 < 4$ (3 OUT OF 4 PARTS).

2. IMPROPER FRACTION:

THE NUMERATOR IS LARGER THAN OR EQUAL TO THE DENOMINATOR AS $X \geq Y$.

FOR EXAMPLE; $\frac{5}{3}, 5 > 3$

3. MIXED FRACTION:

A WHOLE NUMBER AND A FRACTION TOGETHER IS CALLED MIXED FRACTION.

$2\frac{1}{4}$ (2 WHOLE PARTS AND 1 OUT OF 4 PARTS).

4. **UNIT FRACTION**:
HERE THE NUMERATOR IS ALWAYS 1.

$\frac{1}{5}$ (1 PART OF 5 EQUAL PARTS).

5. **LIKE FRACTIONS**:
FRACTIONS WITH THE SAME DENOMINATOR ARE CALLED LIKE FRACTIONS. **FOR EXAMPLE;**

$\frac{4}{7}, \frac{5}{7}$, *are like fractions as their denominator is same is same is same*

UNLIKE FRACTIONS:
FRACTIONS WITH DIFFERENT DENOMINATORS ARE CALLED UNLIKE FRACTIONS.

FOR EXAMPLE;
$\frac{2}{9}, \frac{6}{8}, \frac{5}{3}$ *here denominators are different* 9, 8, 3

6. **EQUIVALENT FRACTIONS**: FRACTIONS THAT LOOK DIFFERENT BUT MEAN THE SAME.

FOR EXAMPLE; $\frac{1}{2}, \frac{2}{4}, \frac{3}{6}$ *here in all the three fraction coefficient is same* 0.5

FOR EXAMPLE; IMAGINE YOU HAVE A PIZZA WITH 8 SLICES.

IF YOU EAT 3 SLICES, YOU ATE $\frac{3}{8}$

IF YOUR FRIEND EATS 5 SLICES, THEY ATE $\frac{5}{8}$

FACE VALUE:

THE FACE VALUE OF A DIGIT IS THE VALUE OF THE DIGIT ITSELF, REGARDLESS OF ITS POSITION IN THE NUMBER. IT DOES NOT DEPEND ON THE PLACE OF THE DIGIT.

FOR EXAMPLE; IN THE NUMBER 572

THE FACE VALUE OF 5 IS 5.

THE FACE VALUE OF 7 IS 7 AND THE FACE VALUE OF 2 IS 2.

PLACE VALUE:

THE PLACE VALUE OF A DIGIT DEPENDS ON ITS POSITION IN THE NUMBER.

IT IS CALCULATED BY MULTIPLYING THE DIGIT BY THE VALUE OF ITS POSITION (ONES, TENS, HUNDREDS, ETC.).

PLACE VALUE POSITIONS:

UNITS/ONES PLACE: VALUE × 1

TENS PLACE: VALUE × 10

HUNDREDS PLACE: VALUE × 100

THOUSANDS PLACE: VALUE × 1000

TEN-THOUSANDS PLACE: VALUE × 10,000 ETC.

FOR EXAMPLE;IN THE NUMBER 7248

THE PLACE VALUE OF 7 IS 7,000 (7 × 1000)

THE PLACE VALUE OF 2 IS 200 (2 × 100)

THE PLACE VALUE OF 4 IS 40 (4 × 10)

THE PLACE VALUE OF 8 IS 8 (8 × 1)

INTEGERS:

POSITIVE AND NEGATIVE NATURAL NUMBERS INCLUDING ZERO ARE CALL INTEGERS.

FOR EXAMPLE;

… -6, -5, -4, -3, -2, -1, 0, +1, +2, +3, +4, +5, +6,…SO ON. HERE -6, -5, -4, -3, -2, -1 REPRESENT NEGATIVE NATURAL NUMBERS AND +1, +2, +3, +4, +5, +6…. REPRESENT POSITIVE NATURAL NUMBERS. ZERO IS ALSO AN INTEGER.

NOTE: INTEGERS DOESN'T REPRESENT NUMBERS WITH DECIMALS, ROOTS, FRACTIONS AS THEY ARE NOT SIMPLE NUMBERS.

FOR EXAMPLE; 1.5, 0.8, $\sqrt{5}$, $\sqrt{3}$ $\frac{5}{8}$, *etc are not integers*

ADDITION:

1) ADDITION OF WHOLE NUMBERS

+5 + 5= +10,   +12 + 5= +17,   +5 + 6= +11

+14 + 16= +30, +150 + 26= +176

+56 + 89= + 145, +1675 + 567 = + 1242

PRACTICE:

+100 + 20= + , +756 + 127= +

+679 + 675 = + , +679 + 456= +

2) ADDITION OF RATIONAL NUMBERS

$$+\frac{1}{3}+\frac{4}{3}=\frac{+3+12}{9}=+\frac{15}{9}$$

$$+\frac{5}{6}+\frac{1}{7}=\frac{+35+6}{42}=+\frac{41}{42}$$

use crossmultiplication method to solve all these questions

$$+\frac{5}{3}+\frac{1}{9}=\frac{+45+3}{27}=+\frac{48}{27}$$

$$+\frac{1}{8}+\frac{5}{2}=\frac{+2+40}{16}=+\frac{42}{16}$$

PRACTICE:

$+\dfrac{1}{7}+\dfrac{2}{8}=$, $\quad +\dfrac{1}{9}+\dfrac{1}{7}=$, $\quad +\dfrac{1}{9}+\dfrac{7}{8}=$

$+\dfrac{1}{8}+\dfrac{1}{8}=$, $\quad +\dfrac{6}{9}+\dfrac{7}{9}=$, $\quad +\dfrac{9}{4}+\dfrac{7}{8}=$

$+\dfrac{6}{9}+\dfrac{7}{8}=$, $\quad +\dfrac{11}{6}+\dfrac{6}{32}=$, $\quad +\dfrac{1}{6}+\dfrac{9}{11}=$

$+\dfrac{7}{9}+\dfrac{8}{6}=$, $\quad +\dfrac{6}{4}+\dfrac{5}{7}=$, $\quad +\dfrac{7}{9}+\dfrac{5}{7}=$

3) SUBTRACTION OF RATIONAL NUMBERS:

$$+\frac{5}{7} - \frac{1}{2} = \frac{+10 - 7}{14} = +\frac{3}{14}$$

$$-\frac{7}{8} + \frac{1}{4} = \frac{-28 + 8}{32} = -\frac{20}{32}$$

$$-\frac{1}{8} - \frac{4}{3} = \frac{-3 - 32}{24} = -\frac{35}{24}$$

PRACTICE:

$$-\frac{6}{7} + \frac{3}{7} = , \quad -\frac{7}{8} - \frac{9}{5} = , \quad -\frac{7}{9} + \frac{4}{5} =$$

Sign remains always of larger digit while performing the Subtraction.

ADDITION OF IRRATIONAL NUMBERS :

while performing addition or subtraction of rational numbers;
1) If the number inside the root is same, then the number outside the roots are added.

17

2) *If the number inside roots are different then the number outside the roots remain as such and no calculation is done.*

BY RULE NO 1:
$+\sqrt{2} + \sqrt{2} = 2\sqrt{2}, \qquad +3\sqrt{3} + 5\sqrt{3} = 8\sqrt{3}$

BY RULE NO 2: $\sqrt{3} + \sqrt{7} = \sqrt{3} + \sqrt{7},$

PRACTICE:
$+4\sqrt{5} + 3\sqrt{5} =, \qquad +5\sqrt{3} + 7\sqrt{3} =$

$+6\sqrt{11} + 8\sqrt{11} =, \qquad +5\sqrt{8} + 3\sqrt{7} =$

$+5\sqrt{45} + 7\sqrt{7} =, \qquad +7\sqrt{6} + 5\sqrt{6} =$

SUBTRACTION OF IRRATIONAL NUMBERS:

$+\sqrt{3}-\sqrt{3}=0, \quad +2\sqrt{5}-\sqrt{5}=+\sqrt{5}, \quad +5\sqrt{7}-3\sqrt{7}=+2\sqrt{7}$

$+6\sqrt{7}-3\sqrt{5} = +6\sqrt{7}-3\sqrt{5}$

HERE RULE 1 AND RULE 2 ALSO FOLLOWS

MULTIPLICATION OF IRRATIONAL NUMBERS :

$\sqrt{6} \times \sqrt{5} = \sqrt{30}$, $3\sqrt{5} \times 5\sqrt{7} = 15\sqrt{35}$

HERE NUMBERS NEED NOT TO BE SAME, FIRST MULTIPLY INSIDE THE ROOT THEN OUTSIDE NUMBERS.

PRACTICE:

$3\sqrt{8} \times 4\sqrt{7} =$, $6\sqrt{7} \times 5\sqrt{8} =$

PRACTICE ABOUT 100 QUESTIONS OF SAME PATTERN

MULTIPLICATION OF RATIONAL NUMBERS:

$\dfrac{3}{2} \times \dfrac{2}{8} = \dfrac{6}{16}$, $\dfrac{7}{5} \times \dfrac{8}{6} = \dfrac{56}{30}$

PRACTICE ABOUT 100 QUESTION OF SAME PATTERN.

ALGEBRAIC EXPRESSION AND IT'S TYPES:

1) ONE VARIABLE ALGEBRAIC EXPRESSION :

AN ALGEBRAIC EXPRESSION CONTAINS A VARIABLE (**X,Y,Z),** CONSTANT(**ANY NUMBER**), COEFFICIENT(**ANY NUMBER**) AND A FUNDAMENTAL OPERATOR(+,-) OR MATHEMATICAL OPERATOR.

CONSIDER THE BELOW ALGEBRAIC EXPRESSION

$ax + b$,

here x, are called variables, $+$ is called

Fundamental Operator, a is called,

coefficient and b is called constant term.

FOR EXAMPLE: CONSIDERTHE FOLLOWING ALGEBRAIC EXPRESSIONS

$3x + 4$, $5x + 7$, $2x + 9$

3, 5, 2 ARE COEFFICIENTS.

X IS VARIABLE.

4, 7, 9 ARE CONSTANT TERMS.

2) TWO VARIABLE ALGEBRAIC EXPRESSION:

AN ALGEBRAIC EXPRESSION CONTAINING TWO VARIABLES (**X AND Y**), COEFFICIENTS, FUNDAMENTAL OPERATOR AND CONSTANT TERM.

$ax + by + c$

here x,y are variables, a,b are

coefficients , + Fundamental operator,

and c constant term.

FOR EXAMPLE:

$3x + 5y + 3,\ 6x + 2y + 8,\ x + y + 1$

ALGEBRAIC EQUATION:

AN ALGEBRAIC EXPRESSION CONTAINING SIGN OF EQUALITY IS CALLED ALGEBRAIC EQUATION:

FOR EXAMPLE:

$ax + b = 0, \quad ax + by + c = 0$

LINEAR EQUATION AND IT'S TYPES:

WHEN IN AN ALGEBRAIC EQUATION, THE VARIABLE HAS HIGHEST POWER AS UNITY OR 1.

1) LINEAR EQUATION IN ONE VARIABLE:

LINEAR EQUATION CONTAINING ONLY ONE VARIABLE **X** OR **Y**.

$ax + b = 0$

2) LINEAR EQUATION IN TWO VARIABLES:

LINEAR EQUATION CONTAINING TWO VARIABLES **X** AS WELL AS **Y**.

$$ax + by + c = 0$$

QUADRATIC EQUATION:

AN ALGEBRAIC EQUATION WHOSE HIGHEST POWER OF VARIABLE IS **2** IS CALLED A QUADRATIC EQUATION:

$ax^2 + bx + c = 0$

CUBIC EQUATION:

AN EQUATION WHOSE HIGHEST POWER OF VARIABLE IS **3** IS CALLED A CUBIC EQUATION:

$ax^3 + bx^2 + cx + d = 0$

POLYNOMIAL AND IT'S TYPES:

THE ALGEBRAIC EXPRESSION CONTAINING MANY TERMS IS CALLED A POLYNOMIAL:

1) MONOMIAL:

ALGEBRAIC EXPRESSION CONTAINING ONLY ONE TERM IS CALLED AS MONOMIAL:

FOR EXAMPLE:

$3,\ 2x,\ 7xy,\ 13x^2,$

2) BINOMIAL:

ALGEBRAIC EXPRESSION CONTAINING TWO TERMS IS CALLED BINOMIAL:

FOR EXAMPLE:

$x + 1,\ 4x + 2,\ 3xy^2 + 6$

3) TRINOMIAL:

ALGEBRAIC EXPRESSION CONTAINING MORE THREE TERMS:

FOR EXAMPLE:

$x + y + z,\qquad 3x^3 + 2x^2 + 3x$

SOME COMMON CONCEPTS FOR KIDS:

1) ANYTHING RAISED POWER ZERO ,IS ALWAYS EQUAL TO 1 , EXCEPT 0:
$x^0 = 1,\ but\ 0^0 \neq 1$

2) **1** IS NOT A PRIME NUMBER

3) **ZERO** IS AN INTEGER

4) **2** IS THE SMALLEST EVEN PRIME NUMBER.

5) ALL NATURAL NUMBERS ARE WHOLE NUMBERS

6) **ZERO** IS DIVISIBLE OR MULTIPLIED BY ANY NUMBER, EQUALS TO ZERO ALWAYS.

BASICS OF HCF AND LCM:

HCF/GCD:

HIGHEST COMMON FACTOR OR GREATEST COMMON DIVISOR
IT CAN BE EASILY CALCULATED BY USING **PRIME FACTORISATION METHOD**
COMMON PRIME FACTORS:

$12 = 2^2 \times 3$

$24 = 2^3 \times 3$,

HERE COMMON PRIME FACTORs are 2 and 3.

HCF IS THE PRODUCT OF LOWEST POWERS OF EACH COMMON PRIME FACTOR:

FOR EXAMPLE:

FIND **HCF** OF **(16, 18)**

STEP 1: FIND PRIME FACTORS OF EACH NUMBER

STEP 2: FIND COMMON PRIME FACTORS OF BOTH THE NUMBERS.

STEP 3: FIND THE PRODUCT OF LOWEST POWERS OF EACH COMMON PRIME FACTOR.

$16 = 2^4$

$18 = 2 \times 3^2$,

HCF= PRODUCT OF LOWEST POWERS OF EACH COMMON PRIME FACTOR= 2

HENCE HCF IS 2

LCM: LEAST COMMON MULTIPLE

STEP 1 : FIND FACTORS OF EACH NUMBER

STEP 2: FIND COMMON FACTORS

STEP 3: FIND THE PRODUCT HIGHEST POWERS OF EACH COMMON PRIME FACTOR

FOR EXAMPLE

FIND LCM OF **(12, 18)**

$12 = 2^3 \times 3$

$18 = 2 \times 3^2$

LCM = PRODUCT OF HEIGHEST POWERS OF EACH PRIME FACTOR $= 2^3 \times 3^2 = 72$,

GEOMETRY:

GEOMETRY IN MATHEMATICS IS THE STUDY OF SHAPES, SIZES, AND SPACES. IT HELPS US UNDERSTAND AND EXPLORE THE WORLD AROUND US BY LEARNING ABOUT THINGS LIKE **LINES**, **CIRCLES**, **TRIANGLES**, AND OTHER **SHAPES**.

SQUARE:

A **SQUARE** IS A FOUR-SIDED SHAPE (QUADRILATERAL) WHERE ALL SIDES ARE EQUAL IN LENGTH AND ALL ANGLES ARE RIGHT ANGLES (90 DEGREES).

1. A SQUARE HAS **FOUR EQUAL SIDES** AND FOUR RIGHT ANGLES.

2. THE DIAGONALS OF A SQUARE ARE EQUAL AND BISECT EACH OTHER AT 90 DEGREES.

3. ALL SQUARES ARE ALSO **RECTANGLES** AND **RHOMBUSES**

4. THE AREA OF A SQUARE IS CALCULATED AS **SIDE × SIDE**.

5. A SQUARE'S PERIMETER IS **4 × SIDE LENGTH.**

6). THE DIAGONALS OF A SQUARE DIVIDE IT INTO FOUR RIGHT TRIANGLES.

7. A SQUARE HAS **FOUR LINES OF SYMMETRY**

8). SQUARES ARE THE ONLY QUADRILATERAL WITH EQUAL SIDES AND ANGLES.

9. THE ANGLES BETWEEN A SQUARE'S SIDES ARE ALWAYS **90 DEGREES.**

10. SQUARES ARE FOUND EVERYWHERE, FROM CHESSBOARDS TO WINDOW PANES!

CUBE:

A **CUBE** IS A 3D SHAPE (SOLID) IN GEOMETRY THAT **HAS SIX EQUAL SQUARE FACES, 12 EQUAL EDGES,** AND **8 VERTICES (CORNERS)**. IT IS A SPECIAL TYPE OF RECTANGULAR PRISM WHERE ALL SIDES ARE THE SAME LENGTH.

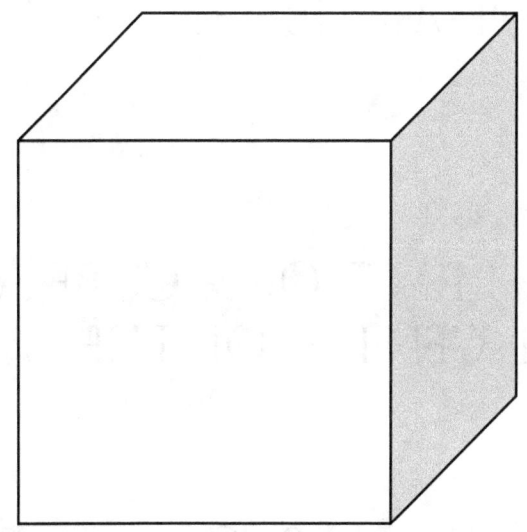

1. A CUBE HAS **6 FACES**, ALL OF WHICH ARE PERFECT SQUARES.

2. IT HAS **12 EDGES**, AND ALL EDGES ARE OF EQUAL LENGTH.

3. A CUBE HAS **8 VERTICES (CORNERS)** WHERE THREE EDGES MEET.

4. THE SURFACE AREA OF A CUBE IS CALCULATED AS **6 × (SIDE²)**.

5. THE VOLUME OF A CUBE IS **SIDE³**.

6. EACH CORNER (VERTEX) OF A CUBE FORMS A **90-DEGREE ANGLE**.

7. THE DIAGONALS ON EACH FACE OF A CUBE ARE EQUAL AND MEET AT THE CENTER OF THE FACE.

8. A CUBE HAS **4 SPACE DIAGONALS** (CONNECTING OPPOSITE CORNERS), AND THEY ARE LONGER THAN THE FACE DIAGONALS.

9. A CUBE IS A **REGULAR HEXAHEDRON**, MEANING IT IS ONE OF THE PLATONIC SOLIDS.

10. EXAMPLES OF CUBES IN REAL LIFE INCLUDE DICE, ICE CUBES, AND RUBIK'S CUBES.

CIRCLE:

A **CIRCLE** IS A 2D SHAPE WHERE ALL POINTS ON ITS BOUNDARY ARE EQUIDISTANT FROM A CENTRAL POINT CALLED THE **CENTER**. THE DISTANCE FROM THE CENTER TO ANY POINT ON THE CIRCLE IS CALLED THE **RADIUS**, AND THE LINE PASSING THROUGH THE CENTER CONNECTING TWO POINTS ON THE CIRCLE IS THE **DIAMETER**.

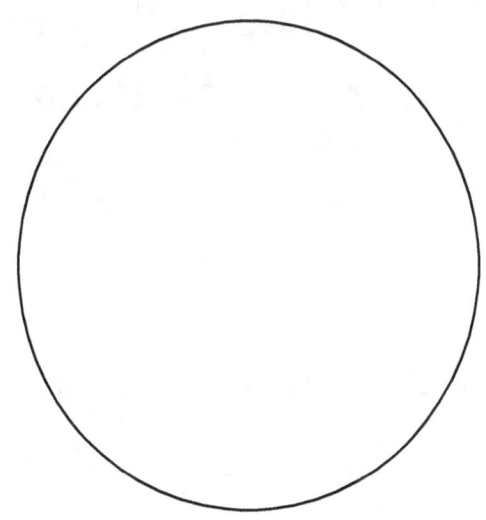

1. THE **RADIUS R** IS HALF THE LENGTH OF THE **DIAMETER D**.

2. THE **CIRCUMFERENCE** IS THE DISTANCE AROUND THE CIRCLE AND IS CALCULATED AS **2πR** (WHERE **R** IS THE RADIUS).

3. THE **AREA** OF A CIRCLE IS GIVEN BY πR^2.

4. THE LONGEST STRAIGHT LINE THAT CAN BE DRAWN IN A CIRCLE IS ITS **DIAMETER**.

5. A CIRCLE HAS **NO CORNERS OR EDGES**; IT IS A CONTINUOUS CURVE.

6. THE RATIO OF A CIRCLE'S CIRCUMFERENCE TO ITS DIAMETER IS ALWAYS π **(APPROXIMATELY 3.14159)**.

7. A LINE THAT TOUCHES A CIRCLE AT EXACTLY ONE POINT IS CALLED A **TANGENT.**

8. A **CHORD** IS A LINE SEGMENT CONNECTING TWO POINTS ON THE CIRCLE; THE LONGEST CHORD IS THE DIAMETER.

9. A CIRCLE IS PERFECTLY **SYMMETRICAL** ABOUT ITS CENTER AND HAS INFINITE LINES OF SYMMETRY.

10. THE **ARC** IS A PORTION OF THE CIRCLE'S BOUNDARY, AND THE ANGLE IT SUBTENDS AT THE CENTER IS CALLED THE CENTRAL ANGLE.

CONE:

A **CONE** IN MATHEMATICS IS A THREE-DIMENSIONAL GEOMETRIC SHAPE THAT HAS A CIRCULAR BASE AND A POINTED TOP, CALLED THE APEX. IT IS FORMED BY CONNECTING ALL THE POINTS ON THE BOUNDARY OF THE CIRCULAR BASE TO THE APEX.

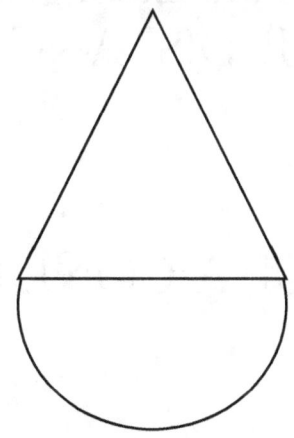

1. **BASIC STRUCTURE**:
   A CONE CONSISTS OF A **CIRCULAR BASE**, A **LATERAL CURVED SURFACE**, AND AN **APEX** (THE POINT WHERE THE LATERAL SURFACE MEETS).

2. **TYPES OF CONES**:

   RIGHT CONE: THE APEX IS DIRECTLY ABOVE THE CENTER OF THE BASE.

OBLIQUE CONE: THE APEX IS NOT ALIGNED VERTICALLY ABOVE THE CENTER.

3. **VOLUME FORMULA**:
   THE VOLUME OF A CONE IS GIVEN BY:

   $$v = \frac{1}{3}\pi R^2 H$$

   WHERE R IS THE RADIUS OF THE BASE, AND H IS THE HEIGHT (PERPENDICULAR DISTANCE FROM THE BASE TO THE APEX).

4. **SURFACE AREA**:
   THE TOTAL SURFACE AREA OF A CONE IS:

   $$A = \pi R(R + L)$$

   WHERE R IS THE RADIUS, AND L IS THE SLANT HEIGHT.

5. **SLANT HEIGHT**:
   THE SLANT HEIGHT (L) CAN BE CALCULATED USING THE PYTHAGOREAN THEOREM:

   $$L = \sqrt{R^2 + H^2}$$

   WHERE H IS THE HEIGHT OF THE CONE.

6. **NET OF A CONE**:
THE 2D NET OF A CONE CONSISTS OF A SECTOR (REPRESENTING THE CURVED SURFACE) AND A CIRCLE (THE BASE).

7. **RELATION TO CYLINDER**:
A CONE'S VOLUME IS ONE-THIRD THE VOLUME OF A CYLINDER WITH THE SAME BASE AREA AND HEIGHT.

8. **REAL-LIFE EXAMPLES**:
COMMON EXAMPLES OF CONES IN REAL LIFE INCLUDE ICE CREAM CONES, TRAFFIC CONES, AND FUNNEL-SHAPED OBJECTS.

9. **AXIS OF SYMMETRY**:
A RIGHT CIRCULAR CONE IS SYMMETRIC ABOUT ITS CENTRAL AXIS (THE LINE PASSING THROUGH THE APEX AND THE CENTER OF THE BASE).

10. **APPLICATIONS**:
CONES ARE USED IN ENGINEERING (CONSTRUCTION OF FUNNELS AND ROCKETS),

CYLINDER:

A **CYLINDER** IN GEOMETRY IS A THREE-DIMENSIONAL SOLID WITH TWO PARALLEL AND CONGRUENT CIRCULAR BASES CONNECTED BY A CURVED SURFACE. THE AXIS OF THE CYLINDER IS THE LINE SEGMENT JOINING THE CENTERS OF THE TWO BASES.

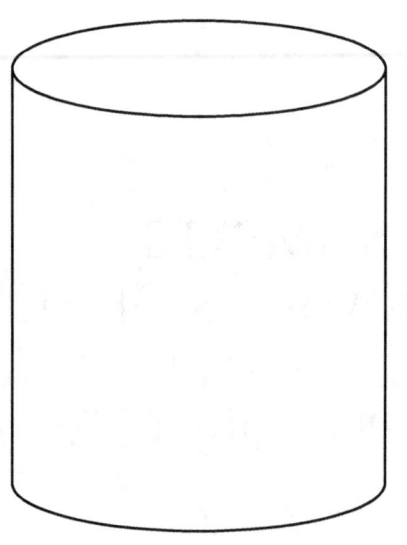

1. **BASIC STRUCTURE**:
A CYLINDER CONSISTS OF:

TWO CIRCULAR BASES (TOP AND BOTTOM)

A CURVED LATERAL SURFACE CONNECTING THE BASES.

2. **TYPES OF CYLINDERS**:

RIGHT CYLINDER: THE AXIS IS PERPENDICULAR TO THE BASES.

OBLIQUE CYLINDER: THE AXIS IS INCLINED AND NOT PERPENDICULAR TO THE BASES.

3. **VOLUME FORMULA**:
   THE VOLUME OF A CYLINDER IS GIVEN BY:

   $V = \pi R^2 H$

   WHERE R IS THE RADIUS OF THE BASE, AND H IS THE HEIGHT.

4. **SURFACE AREA FORMULA**:
   THE TOTAL SURFACE AREA OF A CYLINDER IS:

   $A = 2\pi RH + 2\pi R^2$

   WHERE $2\pi RH$ IS THE LATERAL SURFACE AND $2\pi R^2$ IS THE AREA OF THE TWO CIRCULAR BASES.

5. **AXIS OF SYMMETRY**:
   A RIGHT CYLINDER IS SYMMETRIC ABOUT ITS CENTRAL AXIS, WHICH PASSES THROUGH THE CENTERS OF ITS BASES.

6. **RELATION TO CONE**:
   THE VOLUME OF A CYLINDER IS THREE TIMES THE

VOLUME OF A CONE WITH THE SAME BASE RADIUS AND HEIGHT.

7. **REAL-LIFE EXAMPLES**:
CYLINDERS ARE COMMONLY SEEN IN OBJECTS LIKE SODA CANS, PIPES, BATTERIES, AND STORAGE TANKS.

8. **CROSS-SECTIONS**:

   A VERTICAL CROSS-SECTION OF A CYLINDER IS A RECTANGLE.

   A HORIZONTAL CROSS-SECTION IS A CIRCLE.

9. **NET OF A CYLINDER**:
THE 2D NET OF A CYLINDER CONSISTS OF:

   TWO CIRCLES (THE BASES)

   ONE RECTANGLE (REPRESENTING THE CURVED LATERAL SURFACE).

10. **APPLICATIONS**:
CYLINDERS ARE USED IN MECHANICAL ENGINEERING (PISTON CHAMBERS, PIPES), ARCHITECTURE, AND MANUFACTURING (TANKS AND CONTAINERS).

TRIANGLE:

A **TRIANGLE** IS A POLYGON WITH THREE EDGES AND THREE VERTICES. IT IS ONE OF THE SIMPLEST AND MOST FUNDAMENTAL SHAPES IN GEOMETRY.

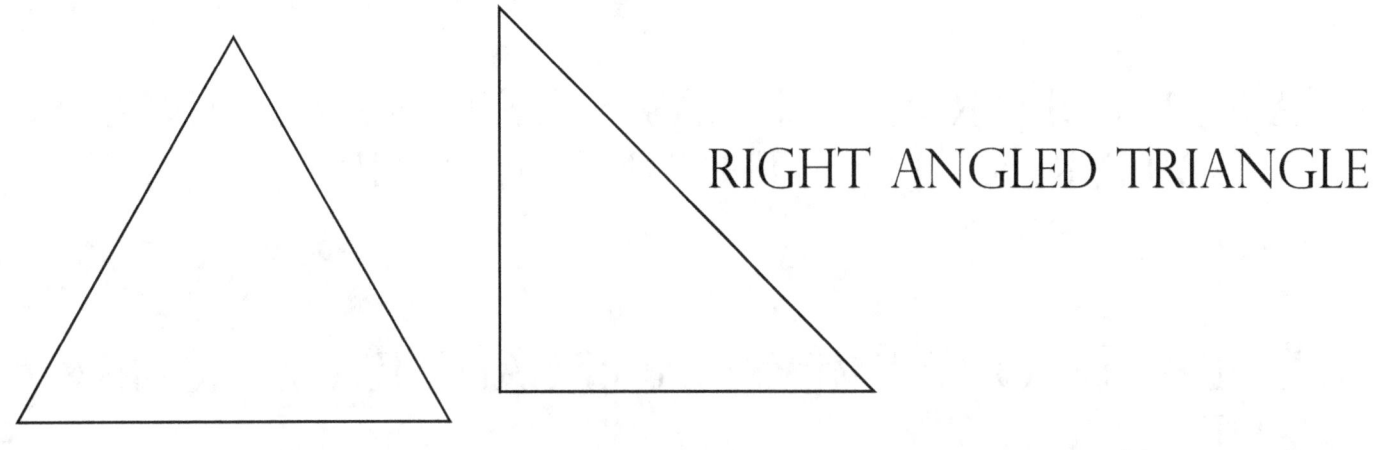

RIGHT ANGLED TRIANGLE

1. A TRIANGLE HAS THREE SIDES, THREE ANGLES, AND THREE VERTICES.

2. THE SUM OF THE INTERIOR ANGLES OF A TRIANGLE IS ALWAYS 180°.

3. AN EQUILATERAL TRIANGLE HAS ALL THREE SIDES AND ANGLES EQUAL (60° EACH).

4. AN ISOSCELES TRIANGLE HAS TWO SIDES OF EQUAL LENGTH AND TWO EQUAL ANGLES.

5. A SCALENE TRIANGLE HAS ALL SIDES AND ANGLES OF DIFFERENT LENGTHS AND MEASURES.

6. THE AREA OF A TRIANGLE IS CALCULATED USING THE FORMULA:

$$A = \frac{1}{2} BASE \times HEIGHT$$

7. THE PYTHAGOREAN THEOREM APPLIES TO RIGHT-ANGLED TRIANGLES:

$$A^2 + B^2 = C^2$$

8. THE LONGEST SIDE IN A RIGHT-ANGLED TRIANGLE IS THE HYPOTENUSE.

9. THE PERIMETER OF A TRIANGLE IS THE SUM OF THE LENGTHS OF ITS SIDES.

10. TRIANGLES ARE CLASSIFIED BASED ON THEIR ANGLES: **ACUTE** (ALL ANGLES < 90°), **RIGHT** (ONE ANGLE = 90°), OR **OBTUSE** (ONE ANGLE > 90°).

SOME INTERESTING FUN FACTS:

FACTS ABOUT ZERO:

HERE ARE SOME FASCINATING FACTS ABOUT **ZERO**:

1. **ZERO IS A CONCEPT**:

ZERO WAS FIRST USED IN ANCIENT CIVILIZATIONS, AND ITS CONCEPT REVOLUTIONIZED MATHEMATICS BY PROVIDING A WAY TO REPRESENT NOTHINGNESS.

2. **ZERO IS NOT A NUMBER IN ANCIENT CULTURES**:

IN MANY EARLY CULTURES, ZERO WAS NOT CONSIDERED A NUMBER. THE ANCIENT GREEKS, FOR EXAMPLE, DID NOT USE ZERO IN THEIR NUMBER SYSTEM.

3. THE SYMBOL FOR ZERO:

THE MODERN SYMBOL FOR ZERO (**0**) WAS DEVELOPED BY INDIAN MATHEMATICIANS IN THE 5TH CENTURY, AND IT SPREAD TO THE WESTERN WORLD THROUGH ARABIC SCHOLARS.

4. ZERO IN OPERATIONS:

ZERO IS THE **ADDITIVE IDENTITY** IN ARITHMETIC, MEANING ANY NUMBER ADDED TO ZERO REMAINS UNCHANGED (E.G., (5 + 0 = 5).

5. MULTIPLYING BY ZERO:

ANY NUMBER MULTIPLIED BY ZERO RESULTS IN ZERO, FOR EXAMPLE, 7×0=0

6. ZERO AS A PLACEHOLDER:

IN THE PLACE-VALUE NUMBER SYSTEM, ZERO IS ESSENTIAL AS A PLACEHOLDER, ALLOWING NUMBERS TO BE WRITTEN EFFICIENTLY (E.G., THE DIFFERENCE BETWEEN 10 AND 100 IS ONLY THE PLACEMENT OF A ZERO).

7. ZERO AND DIVISION:

DIVISION BY ZERO IS UNDEFINED IN MATHEMATICS. IT LEADS TO INDETERMINATE OR INFINITE RESULTS AND IS NOT POSSIBLE IN NORMAL ARITHMETIC.

7. ZERO IN CALCULUS:

IN CALCULUS, ZERO IS ESSENTIAL WHEN CALCULATING LIMITS, DERIVATIVES, AND INTEGRALS. FOR EXAMPLE, THE CONCEPT OF APPROACHING ZERO IS KEY TO DEFINING INSTANTANEOUS RATES OF CHANGE.

8. ZERO AND NEGATIVE NUMBERS:

ZERO ACTS AS THE BOUNDARY BETWEEN POSITIVE AND NEGATIVE NUMBERS ON THE NUMBER LINE.

9. ZERO IN COMPUTERS:

IN BINARY CODE, WHICH IS THE LANGUAGE OF COMPUTERS, ZERO IS ONE OF THE TWO DIGITS USED, REPRESENTING OFF OR NO SIGNAL IN THE SYSTEM.

10. ZERO IN COMPUTERS:

IN BINARY CODE, WHICH IS THE LANGUAGE OF COMPUTERS, ZERO IS ONE OF THE TWO DIGITS USED, REPRESENTING OFF OR NO SIGNAL IN THE SYSTEM.

FACTS ABOUT π:

1. π IS AN IRRATIONAL NUMBER

π CANNOT BE EXPRESSED AS A SIMPLE FRACTION. ITS DECIMAL REPRESENTATION GOES ON FOREVER WITHOUT REPEATING (APPROXIMATELY **3.14159**).

2. π IS TRANSCENDENTAL:

NOT ONLY IS π IRRATIONAL, BUT IT IS ALSO TRANSCENDENTAL, MEANING IT IS NOT THE ROOT OF ANY NON-ZERO POLYNOMIAL EQUATION WITH RATIONAL COEFFICIENTS.

3. π IN CIRCLES:

π IS THE RATIO OF A CIRCLE'S CIRCUMFERENCE TO ITS DIAMETER. THIS RATIO IS THE SAME FOR ALL CIRCLES, REGARDLESS OF SIZE.

4. π DAY:

π DAY IS CELEBRATED ON MARCH 14TH (3/14) SINCE THE FIRST THREE DIGITS OF π ARE **3.14.**

5. π AND THE UNIVERSE:

π APPEARS NOT ONLY IN GEOMETRY BUT ALSO IN PHYSICS, ENGINEERING, AND EVEN IN THE STRUCTURE OF THE UNIVERSE. IT'S INVOLVED IN FORMULAS FOR AREAS, VOLUMES, AND IN MANY FIELDS SUCH AS WAVE THEORY AND RELATIVITY.

6. π'S DECIMAL EXPANSION:

THE DIGITS OF π NEVER END AND NEVER REPEAT. AS OF TODAY, OVER 50 TRILLION DIGITS OF π HAVE

BEEN CALCULATED, AND NO PATTERN HAS BEEN FOUND.

7. THE π SYMBOL:

THE GREEK LETTER π WAS FIRST USED TO REPRESENT THE RATIO OF THE CIRCUMFERENCE TO THE DIAMETER BY WELSH MATHEMATICIAN WILLIAM JONES IN 1706.

8. π IN ANCIENT CIVILIZATIONS:

ANCIENT CIVILIZATIONS, INCLUDING THE BABYLONIANS AND EGYPTIANS, APPROXIMATED π IN DIFFERENT WAYS. THE BABYLONIANS USED 3.125 (A BIT LESS ACCURATE THAN THE MODERN VALUE), AND THE EGYPTIANS USED 3.16.

9. π IN POPULAR CULTURE:

π IS OFTEN REFERENCED IN LITERATURE, MUSIC, AND MOVIES. FOR EXAMPLE, THE MOVIE **LIFE OF π** AND THE BOOK **π: A BIOGRAPHY OF THE WORLD'S MOST FAMOUS NUMBER** BOTH EXPLORE THE NUMBER'S SIGNIFICANCE.

10. π AND COMPUTERS:

π HAS BEEN USED AS A TEST CASE IN COMPUTER SCIENCE. CALCULATING π TO AS MANY DECIMAL PLACES AS POSSIBLE TESTS THE LIMITS OF COMPUTATIONAL POWER.

FACTS ABOUT INFINITY ∞:

1. INFINITY IS NOT A NUMBER:

INFINITY IS A CONCEPT, NOT A SPECIFIC, FINITE NUMBER. IT REPRESENTS AN UNBOUNDED QUANTITY THAT IS LARGER THAN ANY NUMBER, AND IT IS USED TO DESCRIBE THINGS THAT HAVE NO END.

2. DIFFERENT SIZES OF INFINITY:

THERE ARE DIFFERENT "SIZES" OF INFINITY. FOR EXAMPLE, THE SET OF REAL NUMBERS HAS A LARGER INFINITY THAN THE SET OF NATURAL NUMBERS, A CONCEPT INTRODUCED BY GEORG CANTOR.

3. INFINITY IN CALCULUS:

INFINITY IS COMMONLY USED IN CALCULUS WHEN CALCULATING LIMITS. FOR EXAMPLE, AS A FUNCTION APPROACHES A CERTAIN POINT, THE VALUE MAY APPROACH INFINITY, INDICATING IT GROWS WITHOUT BOUND.

4. INFINITY IN GEOMETRY:

IN GEOMETRY, PARALLEL LINES ARE SAID TO MEET AT INFINITY IN PROJECTIVE GEOMETRY. THIS IDEA HELPS TO SIMPLIFY THE CONCEPT OF PERSPECTIVE IN ART AND GEOMETRY.

5. THE INFINITE HOTEL:

THE **HILBERT HOTEL** IS A THOUGHT EXPERIMENT THAT DEMONSTRATES THE COUNTERINTUITIVE PROPERTIES OF INFINITY. EVEN IF THE HOTEL IS FULL, IT CAN STILL ACCOMMODATE MORE GUESTS BY MOVING EACH GUEST TO A NEW ROOM.

6. INFINITY AND THE REAL WORLD:

WHILE INFINITY CAN BE USEFUL IN MATHEMATICAL MODELS, IT DOES NOT EXIST IN THE PHYSICAL WORLD. IT IS A CONCEPT USED TO DESCRIBE IDEALIZED SITUATIONS, SUCH AS AN INFINITE NUMBER OF POINTS ON A LINE.

7. CANTOR'S INFINITY:

GEORG CANTOR PROVED THAT SOME INFINITIES ARE LARGER THAN OTHERS, USING HIS CONCEPT OF **CARDINALITY**. FOR EXAMPLE, THE SET OF REAL NUMBERS IS UNCOUNTABLE INFINITE, WHILE THE SET OF NATURAL NUMBERS IS COUNTABLY INFINITE.

8. INFINITY AND LIMITS::

IN CALCULUS, INFINITY IS OFTEN USED IN THE CONTEXT OF LIMITS. FOR INSTANCE, THE LIMIT OF 1/X AS X APPROACHES 0 IS INFINITY, MEANING IT INCREASES WITHOUT BOUND.

9. INFINITY IN SET THEORY:

IN SET THEORY, THERE ARE INFINITE SETS, LIKE THE SET OF ALL NATURAL NUMBERS, INTEGERS, OR

RATIONAL NUMBERS, EACH WITH DIFFERENT CHARACTERISTICS OF INFINITY.

10. ∞ AS A SYMBOL:

THE SYMBOL FOR INFINITY (∞) WAS FIRST USED BY JOHN WALLIS IN 1655. IT IS BELIEVED TO BE DERIVED FROM THE ROMAN NUMERAL FOR 1000 (WHICH REPRESENTED "A LARGE NUMBER") OR FROM THE LEMNISCATE (A FIGURE-EIGHT SHAPE) SYMBOLIZING ENDLESSNESS.

FACTS ABOUT 1 :

1. THE BUILDING BLOCK OF ALL NUMBERS:

THE NUMBER **1** IS THE BUILDING BLOCK OF ALL OTHER NUMBERS. EVERY OTHER NUMBER CAN BE MADE BY ADDING OR MULTIPLYING **1**.

2. THE MULTIPLICATIVE IDENTITY:

1 IS THE ONLY NUMBER THAT, WHEN MULTIPLIED BY ANY OTHER NUMBER, GIVES THE SAME NUMBER. FOR EXAMPLE, **5×1=5**

3. 1 IS A UNIQUE PRIME:

ALTHOUGH **1** IS NOT CONSIDERED A PRIME NUMBER, IT IS A UNIQUE NUMBER THAT HAS EXACTLY ONE DIVISOR: ITSELF.

4. THE SYMMETRIC IDENTITY:

WHEN YOU ADD **1** TO ANY NUMBER, IT INCREASES THE NUMBER BY EXACTLY ONE. THIS IS THE SIMPLEST FORM OF COUNTING.

5. 1 IN BINARY:

IN THE BINARY SYSTEM, **1** REPRESENTS THE "ON" STATE. IT IS ONE OF ONLY TWO DIGITS USED IN THE BINARY NUMBER SYSTEM (THE OTHER BEING 0).

6. 1 IS THE FIRST ODD NUMBER:

1 IS THE SMALLEST ODD NUMBER IN MATHEMATICS. AN ODD NUMBER IS ANY NUMBER THAT CAN'T BE DIVIDED EVENLY BY 2.

7.1 IN ROMAN NUMERALS:

IN ROMAN NUMERALS, **1** IS REPRESENTED BY THE LETTER 1

8. 1 IS A FACTOR OF EVERY NUMBER:

EVERY NUMBER CAN BE DIVIDED BY **1** WITHOUT CHANGING ITS VALUE. THIS MAKES **1** A FACTOR OF EVERY NUMBER.

9.1 AND THE CONCEPT OF UNITY:

1 IS OFTEN SEEN AS A SYMBOL OF UNITY OR WHOLENESS. IT REPRESENTS A SINGLE ENTITY, WHICH CAN BE POWERFUL IN PHILOSOPHY AND MATHEMATICS.

10.   THE SQUARE ROOT OF 1:

THE SQUARE ROOT OF **1** IS 1. THIS IS BECAUSE 1×1=1 MAKING IT THE ONLY NUMBER WHOSE SQUARE ROOT IS ITSELF.

www.ingramcontent.com/pod-product-compliance
Lightning Source LLC
Chambersburg PA
CBHW082256220526
45469CB00009B/3027